Mastering Mining Engineering Techniques and Innovations for Resource Extraction

Armando

Copyright © [2023]

Title: Mastering Mining Engineering Techniques and Innovations for Resource Extraction
Author's: Armando

This book was printed and published by [Publisher's: **Armando**] in [2023]

ISBN:

TABLE OF CONTENT

Chapter 3: Mining Methods and Techniques

Surface Mining Methods

Open-pit Mining

Strip Mining

Underground Mining Methods

Room and Pillar Mining

Longwall Mining

In-Situ Mining Techniques

Solution Mining

Leaching

Chapter 4: Mine Planning and Design

Preliminary Feasibility Studies

Mine Layout and Design Considerations

Mine Safety and Environmental Factors

Equipment Selection and Optimization

Chapter 5: Mineral Processing and Extraction Techniques 48

Chapter 6: Innovations in Mining Engineering 64

Chapter 10: Career Opportunities and Professional Development in Mining Engineering

Job Roles and Responsibilities in the Mining Industry

Education and Training Pathways for Mining Engineers

Professional Organizations and Networking Opportunities

Chapter 11: Conclusion and Final Remarks

Recap of Key Learnings

The Importance of Mining Engineering in Resource Extraction

Looking Forward: The Future of Mining Engineering

Chapter 1: Introduction to Mining Engineering

The Role and Importance of Mining Engineering

Mining engineering is a critical discipline that plays a pivotal role in the extraction of valuable resources from the Earth's crust. It encompasses a wide range of activities, from identifying potential mining sites to managing the entire mining process. In the book "Mastering Mining Engineering: Techniques and Innovations for Resource Extraction," we delve into the significance of mining engineering and its profound impact on various industries and societies.

One of the primary roles of mining engineering is to locate and assess mineral deposits. Geological engineers, in particular, are responsible for conducting extensive surveys and investigations to identify potential mining sites. By analyzing geological data, they can determine the quantity and quality of mineral resources present in a given area. This information is crucial for mining companies and investors as it helps them make informed decisions about resource extraction.

Furthermore, mining engineering encompasses the design and planning of mining operations. From creating underground tunnels to developing open-pit mines, mining engineers utilize their expertise to design efficient and safe mining methods. They also consider various factors such as environmental impact, safety regulations, and economic viability while planning these operations. This ensures that resources are extracted in a sustainable manner, minimizing any adverse effects on the environment and surrounding communities.

Another key aspect of mining engineering is the management of mining processes. It involves overseeing the extraction, processing, and transportation of minerals. Mining engineers develop strategies to optimize the efficiency of these processes, ensuring maximum resource recovery while minimizing costs. They also implement safety measures to protect the workforce and the environment during mining operations.

The importance of mining engineering extends beyond the mining industry itself. The extraction of minerals is vital for various sectors, including manufacturing, energy production, and construction. Without mining engineers, these industries would struggle to access the raw materials necessary for their operations. Additionally, mining engineering contributes to economic growth by creating job opportunities and generating revenue through the export of minerals.

In conclusion, mining engineering, particularly geological engineering, plays a critical role in the extraction of valuable resources from the Earth's crust. It involves locating and assessing mineral deposits, designing efficient mining methods, and managing the entire mining process. Mining engineering is crucial for various industries and contributes to economic growth. The book "Mastering Mining Engineering: Techniques and Innovations for Resource Extraction" provides valuable insights into this field, making it an essential resource for anyone interested in geological engineering and the broader field of mining engineering.

Historical Overview of Mining Engineering

Throughout history, mining has played a crucial role in the development of civilizations and has been responsible for the extraction of valuable resources that have shaped economies and societies. Mining engineering, as a discipline, has evolved over time to meet the challenges and demands of the industry. This subchapter provides a historical overview of mining engineering, tracing its origins and highlighting key advancements that have shaped the field.

The roots of mining engineering can be traced back to ancient civilizations such as the Egyptians, Greeks, and Romans, who developed various techniques for extracting minerals and metals. These early mining practices laid the foundation for the future development of the field. However, it was during the Industrial Revolution in the 18th and 19th centuries that mining engineering experienced significant advancements and became a distinct discipline.

The Industrial Revolution brought about a rapid increase in the demand for minerals and metals, leading to the need for more efficient and productive mining methods. Engineers played a crucial role in developing innovative techniques and technologies to meet this growing demand. For example, the invention of steam-powered machinery revolutionized mining operations, allowing for deeper and more extensive excavation.

In the late 19th and early 20th centuries, mining engineering continued to evolve with the introduction of new technologies and the establishment of formal education programs. The development of electrical power and the use of explosives greatly improved the efficiency and safety of mining operations. Additionally, the establishment of mining engineering schools and professional organizations further contributed to the growth and professionalization of the field.

In the modern era, mining engineering has become increasingly specialized and interdisciplinary. With the advent of computer-aided design, remote sensing technologies, and advanced modeling techniques, mining engineers are now equipped with powerful tools to optimize resource extraction and minimize environmental impact. Furthermore, the field has expanded to include areas such as geotechnical engineering, environmental engineering, and mine safety.

Today, mining engineering continues to play a vital role in resource extraction, ensuring the sustainable and responsible utilization of Earth's mineral wealth. The discipline is at the forefront of technological advancements and innovations that enable efficient and environmentally conscious mining practices.

In conclusion, the historical overview of mining engineering demonstrates the field's evolution from ancient mining practices to the sophisticated and multidisciplinary discipline it is today. From the contributions of early civilizations to the technological advancements of the Industrial Revolution and the modern era, mining engineering has continuously adapted to meet the challenges of resource extraction. As the field continues to evolve, it holds immense potential for the geological engineering niche as it combines geological knowledge with engineering expertise to extract resources efficiently and sustainably.

Current Challenges and Opportunities in the Field

Geological engineering is a diverse and rapidly evolving field that plays a crucial role in the extraction of valuable resources from the Earth. As we move towards a more sustainable future, it is essential to understand the current challenges and opportunities that exist in this field. In this subchapter, we will explore some of the key issues and exciting prospects that geological engineering professionals are facing today.

One of the foremost challenges in geological engineering is the need to balance resource extraction with environmental conservation. As the global demand for resources continues to rise, there is a growing emphasis on responsible mining practices that minimize the impact on ecosystems and communities. Geological engineers are tasked with finding innovative solutions to extract resources while ensuring minimal disturbance to the environment. This challenge presents an opportunity for professionals to develop and implement sustainable mining techniques that are economically viable and environmentally friendly.

Another significant challenge is the increasing complexity of resource extraction. Many easily accessible deposits have already been depleted, leading to a shift towards extracting resources from more remote and challenging locations. Geological engineers must overcome various technical and logistical obstacles to extract resources from deep underground or under the sea. This presents an opportunity for the development of advanced technologies and techniques, such as autonomous drilling systems, remote monitoring, and real-time data analytics, to optimize extraction processes.

Furthermore, the field of geological engineering faces the challenge of ensuring the safety and well-being of workers in hazardous mining environments. Developing strategies to mitigate risks associated with rockfalls, cave-ins, and other physical hazards is of utmost importance. Additionally, geological engineers must address the health and safety concerns related to exposure to harmful substances such as silica dust and toxic gases. This challenge creates opportunities for professionals to enhance safety protocols, utilize advanced monitoring systems, and improve emergency response procedures.

In conclusion, geological engineering is a dynamic field with numerous challenges and opportunities. As professionals strive to meet the growing demand for resources while minimizing environmental impact, they must embrace technological advancements and sustainable practices. The field of geological engineering holds immense potential for innovation and growth, making it an exciting and rewarding career choice. By addressing the current challenges and seizing the opportunities that arise, geological engineers can contribute to the sustainable development of our planet's resources for generations to come.

Chapter 2: Fundamentals of Mineral Resources

Types and Classification of Mineral Resources

Mineral resources play a crucial role in various industries, from construction to energy production, making them vital for the development and growth of nations. Understanding the different types and classification of mineral resources is essential for geological engineers and anyone interested in the field of mining engineering. This subchapter will provide an overview of the various types of mineral resources and their classification, shedding light on their significance and exploration techniques.

Mineral resources can be broadly categorized into metallic and non-metallic minerals. Metallic minerals include precious metals like gold, silver, and platinum, as well as base metals like copper, lead, and zinc. These minerals are highly valuable due to their excellent conductivity, malleability, and durability, making them essential for manufacturing electronic devices, vehicles, and infrastructure. On the other hand, non-metallic minerals encompass a wide range of resources such as limestone, gypsum, clay, and sand. These minerals are primarily used in the construction industry for producing cement, bricks, and other building materials.

The classification of mineral resources is based on their geological occurrence and economic viability. Reserves refer to economically recoverable mineral deposits that can be extracted profitably using existing technologies. These reserves are further classified into proven, probable, and possible reserves, depending on the level of confidence in their existence and economic feasibility. Additionally, mineral resources can also be categorized as reserves based on their grade, which indicates the concentration of the valuable mineral in the deposit.

Exploration techniques are crucial for identifying and assessing mineral resources. Geological mapping, geophysical surveys, and geochemical analysis are common techniques used to locate and characterize mineral deposits. The integration of advanced technologies like remote sensing, satellite imagery, and geographic information systems (GIS) has revolutionized the exploration process, allowing geological engineers to identify potential deposits more efficiently and accurately.

Understanding the types and classification of mineral resources is vital for sustainable resource extraction and management. Geological engineers play a crucial role in assessing the availability and quality of mineral resources, ensuring their responsible extraction, and minimizing the environmental impact. By mastering the techniques and innovations in mining engineering, professionals in this field contribute to the development of efficient and sustainable resource extraction practices.

In conclusion, mineral resources are the backbone of various industries, and their types and classification are essential for geological engineers and individuals interested in mining engineering. This subchapter provided an overview of metallic and non-metallic minerals, their economic viability, and the exploration techniques used to identify and assess them. By understanding these concepts, professionals in geological engineering can contribute to responsible resource extraction and sustainable development.

Geological Factors Affecting Resource Extraction

Resource extraction plays a crucial role in meeting the ever-increasing global demand for minerals and energy resources. However, the success of any mining operation heavily relies on understanding and managing various geological factors that can significantly impact the extraction process. In this subchapter, we will delve into the key geological factors that influence resource extraction, providing valuable insights for both aspiring and experienced geological engineers.

1. Geological Structures: The presence of faults, fractures, and folds in the Earth's crust can greatly impact resource extraction. These structures can affect the stability of underground mines, alter the flow of groundwater, and influence the distribution of minerals within the deposit. Understanding the structural geology of the site is crucial for designing safe and efficient mining plans.

2. Rock Properties: Different rock types possess varying physical and mechanical properties, such as hardness, strength, and brittleness. These properties influence the choice of mining methods and equipment. For instance, soft or friable rocks may require less energy-intensive extraction techniques, while hard rocks may necessitate drilling and blasting.

3. Mineralogy: The mineral composition of an ore deposit can impact the extraction process. Minerals with high concentrations of valuable elements are usually preferred for mining operations. Additionally, knowledge of the mineralogy helps in selecting appropriate beneficiation techniques to separate the valuable minerals from the gangue material.

4. Hydrogeology: Understanding the movement and behavior of groundwater is crucial in resource extraction. Water can affect the stability of underground mines and the leaching of minerals. Accurate hydrogeological assessments aid in designing proper dewatering systems, preventing flooding or contamination of mining areas.

5. Geochemical Factors: Geochemical characteristics of the deposit can influence the methods employed for extraction. For example, the presence of acidic rock formations can lead to acid mine drainage, causing environmental concerns. Proper management of geochemical factors is essential to minimize the impact on surrounding ecosystems.

6. Geological Hazards: Geological hazards pose significant challenges to resource extraction. These hazards include landslides, subsidence, seismic activity, and volcanic eruptions. Geological engineers must evaluate and manage these risks to ensure the safety of mining operations and personnel.

Understanding and effectively managing these geological factors can significantly enhance resource extraction processes, making them safer, more efficient, and environmentally sustainable. By integrating geological knowledge into mining engineering practices, professionals can optimize the extraction of valuable resources while minimizing the impact on the natural environment.

Whether you are a geological engineering student, a practicing professional, or simply curious about the field, this subchapter provides a comprehensive overview of the geological factors affecting resource extraction. By mastering these concepts, you will be equipped with the necessary tools to contribute to the sustainable development of mining operations and ensure the responsible extraction of Earth's valuable resources.

Exploration Techniques for Identifying Mineral Deposits

As the demand for mineral resources continues to rise, the need for effective exploration techniques to identify mineral deposits becomes increasingly crucial. This subchapter explores the various methods and innovations used in geological engineering to uncover these valuable resources.

One of the fundamental techniques employed in mineral exploration is geological mapping. This involves the systematic study and documentation of rock formations, mineral assemblages, and structural features in an area of interest. By analyzing the characteristics of the surrounding rocks and their relationship to potential mineralization, geologists can gain valuable insights into the presence and extent of mineral deposits.

Geophysical methods also play a significant role in identifying mineral deposits. These techniques utilize various physical properties of rocks, such as their magnetic, electrical, and seismic characteristics, to detect anomalies that may indicate the presence of minerals. For instance, magnetic surveys can identify magnetic minerals associated with certain ore bodies, while electrical resistivity surveys can locate conductive minerals like copper and gold.

Remote sensing technologies have revolutionized the field of mineral exploration. Satellites equipped with hyperspectral imaging sensors can detect unique spectral signatures of minerals on the Earth's surface. By analyzing the reflected light from different wavelengths, geologists can identify specific minerals and even map their distribution over large areas. This non-invasive method provides a cost-effective and time-efficient way to identify potential mineral deposits.

Drilling is another essential technique in mineral exploration. Core drilling involves extracting cylindrical samples of rock from deep below the surface, allowing geologists to analyze the mineralogy and structure of the subsurface. This method provides valuable information on the quality, quantity, and geometry of mineral deposits.

In recent years, advancements in data analytics and machine learning have revolutionized exploration techniques. By integrating vast amounts of geological, geophysical, and remote sensing data, sophisticated algorithms can identify patterns and anomalies that might otherwise go unnoticed. These innovative approaches enable geologists to target specific areas with a higher probability of hosting mineral deposits, ultimately reducing exploration costs and increasing efficiency.

In conclusion, the identification of mineral deposits is a complex and multidisciplinary task in geological engineering. By combining traditional methods such as geological mapping and drilling with modern technologies like remote sensing and data analytics, geologists can maximize their chances of discovering valuable mineral resources. These exploration techniques continue to evolve, driven by technological advancements and the ever-growing demand for minerals in various industries.

Chapter 3: Mining Methods and Techniques

Surface Mining Methods

Surface mining is a widely used method in the field of geological engineering for extracting valuable minerals and ores from the Earth's surface. This subchapter aims to provide an overview of the different surface mining methods employed in the industry, their benefits, and their impact on the environment.

One of the most common surface mining methods is open-pit mining. This technique involves the excavation of large open pits or quarries to extract minerals or rocks. Open-pit mining is preferred when the ore body is located close to the surface, making it economically viable to extract the material. It allows for efficient extraction of minerals and facilitates the use of heavy machinery for mining operations.

Another widely used surface mining method is strip mining. In strip mining, the topsoil and overburden are removed to expose the mineral deposit. This method is particularly suitable for extracting coal and other sedimentary deposits that are horizontally layered. Strip mining is efficient and cost-effective, as it allows for the simultaneous extraction of multiple layers of minerals.

Mountaintop removal mining is another surface mining method used primarily for extracting coal deposits located in mountainous regions. This method involves removing the summit of a mountain to access the coal seams beneath. While this technique is highly efficient in terms of coal extraction, it has significant environmental consequences, including habitat destruction and water pollution.

Surface mining methods also include placer mining, which involves extracting valuable minerals from alluvial deposits such as rivers, beaches, or sand dunes. Placer mining is commonly used for extracting gold, diamonds, and other heavy minerals. This method utilizes water, gravity, and mechanical separation techniques to separate the valuable minerals from the sediment.

It is important to note that while surface mining methods are efficient for resource extraction, they can have adverse environmental impacts. The removal of topsoil and overburden can lead to erosion, habitat destruction, and alteration of landscapes. Mining operations can also generate large volumes of waste materials, which need to be properly managed to minimize their impact on the environment.

In conclusion, surface mining methods play a crucial role in geological engineering for the extraction of valuable minerals and ores. Open-pit mining, strip mining, mountaintop removal mining, and placer mining are some of the commonly employed techniques. While these methods offer efficient resource extraction, it is essential to adopt sustainable practices and mitigate their environmental impact for the long-term preservation of our planet.

Open-pit Mining

Open-pit mining is a widely used method for extracting minerals and resources from the Earth's surface. This subchapter aims to provide a comprehensive understanding of open-pit mining techniques and innovations for individuals interested in geological engineering. Whether you are a student, professional, or simply curious about the field, this section will introduce you to the fundamental concepts and advancements in this mining method.

Open-pit mining refers to the process of excavating minerals or resources from an open pit or surface excavation. It is typically employed when the desired materials are located near the surface, making it cost-effective and efficient. In this subchapter, we will explore the various stages of open-pit mining, including planning, design, production, and reclamation.

The subchapter begins by delving into the initial stages of open-pit mining, where geological and geotechnical surveys are conducted to assess the quality and quantity of the resource. We will discuss the methods used for exploration, sampling, and resource estimation, providing you with a solid foundation for understanding the subsequent stages.

Moving forward, we will explore the planning and design aspects of open-pit mining. This includes determining the optimal pit dimensions, slope stability analysis, and equipment selection. We will also discuss the importance of environmental considerations and mitigating the impact of mining activities on the surrounding ecosystems.

Furthermore, this subchapter will delve into innovations and advancements in open-pit mining techniques. We will explore the use of computer-based modeling and simulation tools, as well as advanced equipment and technologies that enhance operational efficiency and safety. From autonomous haul trucks to real-time monitoring systems, you will gain insight into the cutting-edge developments in the field.

Lastly, we will emphasize the importance of reclamation and mine closure. Open-pit mining operations must be designed with future land use in mind, ensuring that the site can be restored and rehabilitated once mining activities cease. We will discuss the various reclamation techniques and strategies employed to minimize the long-term environmental impact of mining operations.

By the end of this subchapter, you will have a comprehensive understanding of open-pit mining techniques, innovations, and their relevance in geological engineering. Whether you are pursuing a career in this field or simply seeking knowledge on resource extraction, this section will equip you with the necessary tools and information to navigate the world of open-pit mining.

Strip Mining

Strip mining, also known as open-pit mining or surface mining, is a widely used technique in the field of geological engineering for resource extraction. This subchapter will provide an overview of strip mining, its techniques, and innovations that have transformed the industry.

Strip mining involves the removal of large strips of overburden or vegetation to expose valuable mineral deposits beneath the Earth's surface. It is particularly suitable for extracting shallow deposits of coal, limestone, phosphate, and other minerals. The process begins with the clearing of vegetation and topsoil, followed by the excavation of the mineral deposit.

One of the key advantages of strip mining is its cost-effectiveness. As compared to underground mining, strip mining requires fewer human resources and can access larger areas of mineral deposits. Additionally, the extracted minerals are readily accessible, enabling efficient transportation and processing.

However, strip mining also poses several environmental challenges. The removal of topsoil and vegetation can lead to soil erosion, habitat destruction, and alteration of natural landscapes. To mitigate these impacts, innovative reclamation techniques have been developed. These include the re-establishment of vegetation, contouring of landforms, and restoration of water bodies, ensuring that the land is restored to a functional state after mining operations have ceased.

Technological advancements have played a significant role in improving the efficiency and sustainability of strip mining. For instance, the use of advanced machinery and automation has increased productivity while reducing human involvement in hazardous tasks. Remote sensing technologies, such as LiDAR and satellite imagery, enable precise mapping of the mining area, aiding in planning and monitoring operations.

In recent years, the integration of digital technologies and data analytics has revolutionized strip mining practices. Real-time monitoring systems can track equipment performance, optimize production, and ensure worker safety. Advanced modeling and simulation tools allow engineers to predict the behavior of the mining site, optimizing the extraction process and minimizing environmental impacts.

In conclusion, strip mining is a widely utilized technique in geological engineering for resource extraction. While it offers significant advantages in terms of cost and accessibility, it also presents environmental challenges. Nevertheless, through innovative reclamation techniques and advancements in technology, the industry strives to minimize its ecological footprint and enhance the sustainability of strip mining operations.

Underground Mining Methods

Underground mining methods play a significant role in the extraction of valuable resources from beneath the Earth's surface. This subchapter delves into the techniques and innovations employed in this field, providing valuable insights for professionals and enthusiasts in geological engineering.

Underground mining involves accessing mineral deposits that are buried deep beneath the surface. It is often chosen when the resource is located at a depth that makes surface extraction impractical or uneconomical. This method offers several advantages over open-pit mining, including increased safety, reduced environmental impact, and the ability to access deeper deposits.

The subchapter begins by exploring the various types of underground mining methods commonly employed in the industry. These include room and pillar mining, longwall mining, and block caving. Each method is explained in detail, highlighting their specific advantages, challenges, and applications. The audience gains a comprehensive understanding of the principles behind these techniques and how they contribute to effective resource extraction.

In addition to traditional mining methods, technological advancements and innovations are also covered. This includes the use of automation, robotics, and remote-controlled machinery, which have revolutionized underground mining operations. The audience will learn about the benefits of these technologies, such as increased productivity, improved safety, and enhanced efficiency.

The subchapter also addresses the importance of geotechnical considerations in underground mining. Geological engineers play a crucial role in assessing the stability of the rock mass and designing support systems to prevent collapses and ensure worker safety. The audience will gain insights into the key factors that geological engineers must consider, such as rock strength, stress distribution, and groundwater conditions.

Furthermore, the subchapter explores the environmental impact of underground mining and the measures taken to mitigate its effects. Topics such as mine ventilation, dust control, and water management are discussed to highlight the industry's commitment to sustainable practices.

Overall, this subchapter on underground mining methods provides a comprehensive overview of the techniques and innovations used in geological engineering. It caters to a wide audience, including professionals in the field, students, and anyone interested in understanding the intricacies of resource extraction from beneath the Earth's surface.

Room and Pillar Mining

Room and pillar mining is a widely used technique in the field of geological engineering for resource extraction. This subchapter explores the principles and methods of room and pillar mining, providing valuable insights into this essential process.

Room and pillar mining is particularly suitable for extracting minerals and ores that are found in horizontal deposits. It involves the creation of a series of rooms or chambers within the deposit, leaving pillars of untouched material to support the roof. These pillars act as a safety measure, preventing the collapse of the mined area.

This technique is highly efficient and allows for the extraction of a significant amount of resources. It offers an excellent balance between safety, productivity, and cost-effectiveness. Room and pillar mining is especially favored in coal mining operations, where it is commonly used due to the nature of coal deposits.

This subchapter delves into the various stages of room and pillar mining, from initial planning to the extraction process. It provides a step-by-step guide, detailing the necessary considerations for successful implementation. Readers will gain a comprehensive understanding of the key factors that must be taken into account, such as the strength of the surrounding rock and the size and shape of the pillars.

Furthermore, this subchapter explores the different variations of room and pillar mining, including the classic method and newer innovative approaches. These variations adapt the technique to specific geological conditions, aiming to optimize resource extraction and minimize environmental impact.

Readers will also learn about the challenges and risks associated with room and pillar mining. These include potential roof collapses, pillar instability, and ventilation issues. The subchapter offers advice on how to mitigate these risks effectively, ensuring the safety of workers and the longevity of the mine.

In conclusion, the subchapter on room and pillar mining provides an indispensable resource for anyone interested in geological engineering, especially those involved in resource extraction. It offers a comprehensive overview of this widely used technique, exploring its principles, methods, variations, and associated risks. Whether one is a student, a professional in the field, or simply curious about the world of mining, this subchapter is a valuable addition to their knowledge base.

Longwall Mining

Longwall mining is a highly efficient and widely used technique in the field of geological engineering. This subchapter aims to provide a comprehensive overview of longwall mining, discussing its principles, techniques, and innovations that have revolutionized resource extraction.

Longwall mining is a method of underground coal mining that involves the extraction of coal panels in a continuous process. It differs from traditional mining methods, such as room and pillar mining, as it allows for the complete extraction of the coal seam, resulting in higher productivity and recovery rates. This technique is particularly suitable for large-scale operations, where the coal seam is thick and continuous.

The key concept behind longwall mining is the use of a longwall shearer, a machine equipped with cutting drums that extract coal in a continuous process. As the shearer moves along the coal face, it cuts and loads the coal onto an armored face conveyor. Simultaneously, hydraulic powered supports, known as shields, advance forward to provide support to the roof, preventing cave-ins.

One of the major advantages of longwall mining is its efficiency and productivity. The continuous extraction process allows for higher coal recovery rates compared to traditional methods. Additionally, the mechanized nature of longwall mining reduces the need for manual labor, resulting in safer working conditions for miners.

In recent years, several innovations have been introduced to further enhance the efficiency and sustainability of longwall mining. These include the use of advanced automation systems, high-capacity longwall equipment, and sophisticated monitoring and control technologies. These innovations have not only increased productivity but have also reduced the environmental impact of mining operations.

However, longwall mining also presents challenges and risks that need to be addressed for successful implementation. These include subsidence, ventilation requirements, and safety hazards associated with the use of heavy machinery underground. Proper planning, monitoring, and risk management strategies are crucial to mitigate these challenges and ensure the safety of workers and the surrounding environment.

In conclusion, longwall mining is a highly efficient technique that has revolutionized resource extraction in the field of geological engineering. Its continuous extraction process, along with advancements in technology and innovation, has significantly improved productivity and safety in underground mining operations. As the mining industry continues to evolve, longwall mining techniques will continue to play a vital role in meeting the global demand for natural resources.

In-Situ Mining Techniques

In the field of mining engineering, one of the most innovative and efficient methods of resource extraction is known as in-situ mining. This technique has revolutionized the way we access valuable minerals and resources, particularly in the realm of geological engineering. In this subchapter, we will delve into the intricacies of in-situ mining techniques, exploring their applications, benefits, and potential challenges.

In-situ mining, also referred to as in-situ recovery or solution mining, involves the extraction of minerals or resources directly from their natural location without the need for extensive excavation. This method is especially advantageous when the resource is located deep underground, making traditional mining techniques cumbersome and costly. By implementing in-situ mining, geological engineers can access resources that were previously inaccessible, thereby maximizing the utilization of our planet's rich mineral deposits.

One of the primary advantages of in-situ mining is its minimal environmental impact. Unlike conventional mining, which often results in massive disruptions to the surrounding ecosystem, in-situ mining techniques have a significantly smaller footprint. This leads to reduced land disturbance, preservation of natural habitats, and a lower risk of soil erosion and water contamination. Furthermore, in-situ mining has the potential to minimize greenhouse gas emissions associated with transportation and excavation, making it a more sustainable alternative.

In-situ mining techniques can be broadly classified into two categories: chemical and thermal methods. Chemical methods involve the injection of specific solvents into the resource deposit, which dissolve the target minerals. The resulting solution is then pumped to the surface for further processing. On the other hand, thermal methods rely on heating the resource deposit, causing the minerals to become more soluble or vaporize, enabling their extraction.

While in-situ mining offers numerous benefits, it is not without its challenges. The success of this technique relies heavily on accurate geological characterization, as variations in rock properties can impact the efficiency of mineral extraction. Additionally, the choice of solvents or heating methods must be carefully considered to ensure optimal results and minimize potential environmental risks.

In conclusion, in-situ mining techniques have emerged as a game-changer in the field of mining engineering, particularly in geological engineering. By allowing for the extraction of resources directly from their natural location, in-situ mining minimizes environmental impact, maximizes resource recovery, and offers a more sustainable approach to resource extraction. As we continue to advance in our understanding and implementation of in-situ mining, the potential for unlocking previously untapped mineral deposits becomes increasingly promising, paving the way for a more efficient and responsible approach to resource extraction.

Solution Mining

Solution mining is a highly innovative and efficient technique used in the field of geological engineering for resource extraction. It involves the extraction of valuable minerals or resources from underground deposits by dissolving them in a liquid solution. This subchapter aims to provide an overview of solution mining techniques and highlight the innovations that have revolutionized this field.

One of the key advantages of solution mining is its versatility. It can be applied to various types of deposits, including salt, potash, uranium, and even petroleum. The process begins by drilling wells into the deposit, typically using advanced drilling techniques such as directional drilling. Once the wells are in place, a solvent is injected into the deposit, which dissolves the desired minerals or resources.

In recent years, several innovative techniques and technologies have been developed to enhance the efficiency and sustainability of solution mining. For instance, the use of advanced computer modeling and simulation tools enables engineers to accurately predict the behavior of the solvent and optimize the extraction process. This not only reduces costs but also minimizes the environmental impact.

Another significant innovation in solution mining is the introduction of in-situ recovery (ISR) techniques. ISR involves the injection of a leaching solution into the deposit, which dissolves the minerals or resources. The solution is then pumped back to the surface, where the valuable components are extracted. This method eliminates the need for large-scale excavation and reduces the surface disturbance associated with traditional mining methods.

Moreover, advancements in automation and robotics have played a crucial role in improving the safety and efficiency of solution mining operations. Remote-controlled vehicles and drones are now widely used for monitoring and inspecting underground wells, reducing the risk to workers and increasing productivity.

In conclusion, solution mining is a highly effective technique for resource extraction in the field of geological engineering. It offers numerous advantages, including versatility, efficiency, and sustainability. With continuous innovations and advancements in technology, solution mining is expected to play an increasingly vital role in meeting the world's growing demand for valuable minerals and resources.

Leaching

Leaching is a fundamental process in the field of geological engineering that plays a crucial role in the extraction of valuable resources from the Earth's crust. In this subchapter, we will delve into the intricacies of leaching, exploring its various techniques and innovations that have revolutionized resource extraction in the mining industry.

Leaching can be defined as the process of extracting soluble substances from solid materials by dissolving them in a liquid medium. It is commonly used to extract metals, such as gold, copper, and uranium, from their respective ores. The success of leaching greatly depends on the choice of leaching agents, the optimization of operating conditions, and the proper design of leaching systems.

One of the key innovations in leaching is the use of chemical agents known as leaching reagents. These reagents can be acidic, alkaline, or neutral, and they play a vital role in breaking down the ore and dissolving the target metals. Additionally, advancements in leaching technology have led to the development of innovative techniques such as heap leaching, bioleaching, and pressure leaching.

Heap leaching involves stacking the ore in a heap and then applying a leaching agent to the top of the heap. As the leach solution percolates through the heap, it dissolves the target metals, which are then collected and further processed. This technique is particularly suitable for low-grade ores and has significantly reduced the cost of resource extraction.

Bioleaching is another innovative technique that utilizes microorganisms to extract metals from ores. Certain bacteria and fungi have the ability to oxidize the minerals present in the ore, making them more soluble and accessible for leaching. This environmentally friendly approach has gained popularity due to its lower energy requirements and reduced environmental impact.

Pressure leaching is a technique that involves subjecting the ore to high temperature and pressure conditions, enhancing the leaching process. This method is commonly used for extracting metals from sulfide ores and has proven to be highly efficient in recovering valuable resources.

In conclusion, leaching is a critical process in geological engineering that has revolutionized resource extraction in the mining industry. With advancements in leaching techniques and innovations such as heap leaching, bioleaching, and pressure leaching, the extraction of metals from ores has become more efficient, cost-effective, and environmentally friendly. By mastering the art of leaching, mining engineers can unlock the hidden potential of Earth's crust and ensure the sustainable utilization of its valuable resources.

Chapter 4: Mine Planning and Design

Preliminary Feasibility Studies

In the world of mining engineering, preliminary feasibility studies play a crucial role in determining the viability of a mining project. Before any extraction can take place, it is essential to thoroughly evaluate the geological and engineering aspects to ensure the project's success. This subchapter will delve into the importance of preliminary feasibility studies and the techniques and innovations employed in the field of geological engineering.

Preliminary feasibility studies are the initial investigations conducted by mining engineers to assess the potential of a mining project. These studies involve a comprehensive analysis of several factors, such as the geology of the area, mineral reserves, environmental impact, infrastructure requirements, and financial feasibility. By conducting these studies, engineers can determine whether a project is economically viable and environmentally sustainable.

One of the primary goals of preliminary feasibility studies is to evaluate the geological aspects of a mining project. Geological engineers utilize various techniques to understand the composition of the earth's crust, identify mineral deposits, and estimate their quantity and quality. These techniques include geological mapping, geophysical surveys, drilling, and sampling. By analyzing the obtained data, engineers can determine the potential profitability of a mining venture.

Moreover, preliminary feasibility studies also consider the engineering aspects of a mining project. This includes assessing the infrastructure requirements, such as access roads, power supply, and water sources, as well as the necessary equipment and machinery. Engineers also evaluate the potential environmental impact of the project, with a focus on minimizing any adverse effects on the surrounding ecosystem.

Advancements in technology have revolutionized the field of geological engineering, enabling more accurate and efficient preliminary feasibility studies. Innovations such as remote sensing, satellite imagery, and Geographic Information System (GIS) have made it easier to gather data and analyze it in a comprehensive manner. Additionally, computer modeling and simulation tools allow engineers to simulate different scenarios and optimize the mining operation's efficiency.

In conclusion, preliminary feasibility studies are an integral part of the mining engineering process. By conducting these studies, professionals in the field of geological engineering can assess the viability of a mining project, considering both its geological and engineering aspects. Technological advancements have made these studies more efficient and accurate, ensuring that mining ventures are economically viable and environmentally sustainable.

Mine Layout and Design Considerations

Introduction:

The layout and design of a mine play a crucial role in ensuring safe and efficient resource extraction. Geological engineering encompasses the knowledge and techniques required to plan and execute mining operations effectively. This subchapter aims to discuss the various factors and considerations involved in mine layout and design, providing valuable insights for professionals and enthusiasts interested in the field of geological engineering.

1.   Geotechnical   Considerations:
One of the primary concerns in mine layout and design is the geotechnical aspect. Understanding the geological characteristics of the site, including rock strength, stability, and the presence of faults or fractures, is crucial for determining the optimal mine layout. Geotechnical surveys and assessments help identify potential risks and develop appropriate mitigation measures to ensure safe mining operations.

2.   Access   and   Infrastructure:
Efficient mine access is vital for the transportation of personnel, equipment, and extracted resources. The design of roads, tunnels, ramps, and shafts must take into account the topography, geology, and potential environmental impacts. Additionally, the layout should consider the placement of infrastructure such as workshops, processing plants, and storage facilities to optimize workflow and minimize costs.

3.   Safety   and   Emergency   Preparedness:
Safety is of utmost importance in any mining operation. The layout and design should incorporate safety measures such as proper

ventilation, emergency exits, and fire suppression systems. Adequate escape routes and refuge chambers must be planned to ensure the well-being of workers in the event of an emergency. Additionally, the layout should facilitate easy access for emergency response teams and equipment.

4.   Environmental   Considerations: Mining activities can have significant environmental impacts, which should be minimized and mitigated. The mine layout and design should consider the preservation of ecosystems, water resources, and biodiversity. Implementing measures such as waste management systems, reclamation plans, and sustainable practices can help reduce the ecological footprint of mining operations.

5.   Operational   Efficiency: Efficient mine layout and design can optimize production and reduce operational costs. Proper planning of haulage routes, positioning of equipment, and the arrangement of work areas can enhance productivity. Additionally, incorporating automation and advanced technologies in the design can improve operational efficiency, safety, and resource utilization.

Conclusion:
The layout and design of a mine are critical factors that impact the success, safety, and sustainability of mining operations. Geological engineering professionals must consider various aspects such as geotechnical factors, access and infrastructure, safety measures, environmental considerations, and operational efficiency. By implementing these considerations, mining operations can be conducted responsibly, ensuring the effective extraction of resources

while minimizing environmental impacts and prioritizing worker safety.

Mine Safety and Environmental Factors

In the field of geological engineering, mine safety and environmental factors play a crucial role in ensuring sustainable resource extraction. The extraction of minerals and resources from the Earth is a complex process that requires careful consideration of safety measures and environmental impacts. This subchapter aims to shed light on the importance of mine safety and the various environmental factors that need to be taken into account in the mining industry.

Safety in mining operations is of paramount importance as it directly affects the well-being of the workers and the surrounding communities. Mining sites can be hazardous due to a multitude of factors, such as unstable geological formations, machinery accidents, and exposure to harmful substances. Therefore, it is essential to implement comprehensive safety protocols and training programs to minimize risks and prevent accidents. This subchapter will delve into the specific safety measures and technologies employed in mining operations, including personal protective equipment, monitoring systems, and emergency response procedures.

Furthermore, mining activities can have significant environmental impacts that need to be carefully managed. The extraction of minerals often involves the removal of large amounts of soil and rock, leading to soil erosion and habitat destruction. Additionally, the use of chemicals and the release of pollutants can contaminate water bodies and negatively impact local ecosystems. This subchapter will explore the various environmental management strategies employed in the mining industry, such as waste management, land reclamation, and water treatment techniques. It will also discuss the importance of conducting environmental impact assessments and complying with relevant regulations to ensure responsible mining practices.

Moreover, the subchapter will highlight the advancements and innovations in mine safety and environmental technologies. The integration of automation, robotics, and artificial intelligence has revolutionized the mining industry, making it safer and more efficient. The use of drones for mapping and monitoring purposes has also proved to be a valuable tool in ensuring safety and minimizing environmental impacts.

In conclusion, mine safety and environmental factors are integral aspects of geological engineering. By prioritizing safety measures and implementing sustainable practices, the mining industry can contribute to the responsible extraction of resources while minimizing harm to workers and the environment. This subchapter aims to provide valuable insights into the challenges and advancements in mine safety and environmental management, catering to a wide audience interested in geological engineering and the sustainable development of the mining industry.

Equipment Selection and Optimization

In the field of geological engineering, equipment selection and optimization play a crucial role in ensuring efficient and cost-effective resource extraction. This subchapter aims to provide a comprehensive understanding of the various factors involved in equipment selection and how optimization techniques can enhance mining operations.

The selection of equipment for mining activities is a complex process that requires careful consideration of several factors. One of the primary considerations is the nature of the resource being extracted. Different minerals and ores require specific equipment for extraction and processing. For example, underground mining operations may require specialized machinery such as longwall shearers or continuous miners, while open-pit mining may necessitate the use of large haul trucks and excavators.

Cost is another critical factor in equipment selection. Mining companies need to evaluate the initial capital cost of equipment, as well as the ongoing operating and maintenance expenses. It is essential to strike a balance between upfront investment and long-term operational efficiency to ensure profitability.

Additionally, the environment in which the mining operation takes place must be taken into account. Factors such as terrain, climate, and accessibility can significantly impact the selection of equipment. For example, operations in remote or challenging terrains may require equipment with enhanced mobility and durability.

Once the equipment is selected, optimization techniques can be applied to enhance its performance and productivity. These techniques involve analyzing and fine-tuning various aspects of equipment operation, such as fuel consumption, cycle time, and maintenance schedules.

Fuel consumption optimization is achieved through efficient engine management, reducing idle time, and implementing advanced technologies such as hybrid systems or electric power. This not only reduces operating costs but also contributes to environmental sustainability by minimizing carbon emissions.

Cycle time optimization focuses on improving the efficiency of equipment during the extraction process. Techniques such as improved blasting methods, optimized loading and hauling practices, and real-time monitoring can significantly enhance productivity and reduce downtime.

Furthermore, maintenance optimization is crucial for maximizing equipment availability. Regular inspections, preventive maintenance, and condition monitoring techniques can help identify potential issues before they escalate into major breakdowns. Implementing predictive maintenance strategies using data analytics and sensor technologies can further enhance equipment reliability and reduce unplanned downtime.

In conclusion, equipment selection and optimization are crucial aspects of geological engineering. By considering factors such as resource characteristics, cost, and environmental conditions, mining companies can select the most suitable equipment for their operations. Additionally, optimization techniques can be applied to improve fuel consumption, cycle time, and maintenance practices, leading to enhanced productivity and profitability. This subchapter serves as a valuable guide for anyone involved in the field of geological engineering, providing insights into the intricacies of equipment selection and optimization for successful resource extraction.

Chapter 5: Mineral Processing and Extraction Techniques

Crushing, Grinding, and Sizing

In the world of mining engineering, the processes of crushing, grinding, and sizing are fundamental to the extraction of valuable resources from the earth. These techniques play a crucial role in breaking down rocks and minerals into smaller, more manageable pieces, allowing for efficient mineral separation and extraction.

Crushing is the initial step in the comminution process, where large rocks and ores are reduced in size to facilitate further processing. It involves applying mechanical force to break down the materials into smaller fragments. Various types of crushing equipment, such as jaw crushers, cone crushers, and impact crushers, are employed to accomplish this task. Each type of crusher operates on different principles, but the common goal is to reduce the size of the raw material to a size suitable for further processing.

Once the materials have been crushed, they are ready for the grinding stage. Grinding is a process that uses abrasive particles to reduce the size of the crushed material even further. This step is crucial for liberating valuable minerals from the surrounding gangue, or waste material. Grinding can be performed using various types of equipment, including ball mills, rod mills, and autogenous mills. The choice of grinding equipment depends on factors such as the hardness and size of the material being processed.

Finally, sizing is the process of separating particles based on their size. This step is vital for achieving the desired product size distribution and ensuring that valuable minerals are properly recovered. Various techniques, such as screens, classifiers, and hydrocyclones, are employed for sizing purposes. These devices utilize principles of particle size separation and classification to separate the crushed and ground materials into different size fractions.

In conclusion, crushing, grinding, and sizing are integral processes in mining engineering that are essential for the extraction of valuable resources. These techniques enable the reduction of large rocks and ores into smaller, more manageable pieces, facilitating subsequent mineral separation and extraction. By understanding and mastering these processes, mining engineers can optimize the efficiency and effectiveness of resource extraction, leading to sustainable and responsible mining practices.

This subchapter is intended for a broad audience, including geological engineering professionals and enthusiasts. It serves as a comprehensive introduction to the fundamental processes involved in crushing, grinding, and sizing. By providing a clear understanding of these techniques, readers will gain valuable insights into the world of mining engineering and the critical role it plays in resource extraction.

Physical Separation Techniques

In the field of mining engineering, physical separation techniques play a crucial role in the extraction and processing of valuable resources. These techniques are used to separate and concentrate minerals or metals from the surrounding ore, enabling efficient resource extraction and maximizing the economic viability of mining operations.

Physical separation techniques rely on the differences in physical properties of minerals or metals, such as density, magnetic susceptibility, electrical conductivity, and surface properties. By exploiting these differences, engineers can selectively separate valuable minerals or metals from the ore, making it easier to extract and refine them.

One of the most commonly used physical separation techniques in mining engineering is gravity separation. This technique takes advantage of the differences in particle density to separate minerals or metals of different densities. Gravity separation methods include jigging, heavy media separation, and spiral concentration, among others. These methods are particularly effective for separating dense minerals like gold, platinum, and diamonds.

Another widely used physical separation technique is magnetic separation. This technique utilizes the magnetic properties of certain minerals or metals to separate them from the ore. By applying a magnetic field, engineers can attract and separate magnetic minerals or metals from non-magnetic ones. Magnetic separation is commonly used in iron ore beneficiation and the recovery of valuable metals like nickel and cobalt.

Electrostatic separation is another effective physical separation technique used in mining engineering. This technique relies on the differences in electrical conductivity or surface charge of minerals or metals to separate them. By applying an electric field, engineers can induce charges on particles and separate them based on their response to the field. Electrostatic separation is often used in the recovery of valuable metals like copper, zinc, and lead.

Flotation, a technique based on the differences in surface properties of minerals or metals, is also commonly used in mining engineering. In flotation, minerals or metals are selectively attached to air bubbles, which are then separated from the rest of the slurry. This technique is widely used in the recovery of valuable metals like copper, lead, and zinc.

In conclusion, physical separation techniques are vital tools in the field of mining engineering. They enable the selective separation and concentration of valuable minerals or metals from the surrounding ore, increasing the efficiency and profitability of mining operations. Gravity separation, magnetic separation, electrostatic separation, and flotation are just a few examples of the wide range of physical separation techniques used in the industry. By mastering these techniques, geological engineers can contribute to the sustainable and responsible extraction of valuable resources.

Gravity Separation

Gravity separation is a fundamental technique used in the field of geological engineering for the extraction of valuable resources from ores and minerals. This technique relies on the natural force of gravity to separate different components based on their specific gravity. It is a cost-effective and environmentally friendly method that has been widely adopted in the mining industry.

The principle behind gravity separation is quite simple. Different particles have different densities, which affects their buoyancy in a fluid medium. By subjecting the ore or mineral mixture to a gravitational field, heavier particles will sink to the bottom while lighter particles will rise to the top. This enables the separation of valuable minerals from the gangue, thus enhancing the overall ore quality.

One of the most common gravity separation methods is known as dense media separation (DMS). In DMS, a dense medium such as a suspension of finely ground ferrosilicon or magnetite is used to create a specific gravity close to that of the valuable mineral. The ore is then mixed with the dense medium, and the particles with higher density will sink while the lighter particles will float. This allows for the efficient separation of minerals with different densities.

Gravity separation is particularly useful for the extraction of heavy minerals such as gold, tin, and diamonds. These minerals tend to have a higher specific gravity compared to the surrounding gangue minerals, making them easier to separate using gravity-based methods. Additionally, gravity separation can be applied to recover valuable minerals from low-grade ores or tailings, thereby increasing the overall resource utilization.

Advancements in gravity separation techniques have led to the development of sophisticated equipment and processes. For example, centrifugal gravity separators such as the Knelson concentrator and Falcon concentrator have revolutionized the field by enabling the recovery of fine particles that were previously considered uneconomical. These devices utilize centrifugal force to enhance the gravitational separation process, resulting in higher efficiency and improved recovery rates.

In conclusion, gravity separation is a vital technique in the field of geological engineering for resource extraction. Its simplicity, cost-effectiveness, and environmental friendliness make it a preferred method in the mining industry. With ongoing advancements and innovations, gravity separation continues to play a significant role in maximizing the extraction of valuable minerals from ores and minerals, ultimately contributing to the sustainable utilization of Earth's resources.

Magnetic Separation

In the field of resource extraction, one of the most crucial processes is the separation of valuable minerals from waste materials. Magnetic separation is a technique widely used in the mining industry to achieve this objective. This subchapter will explore the principles behind magnetic separation and its applications in geological engineering.

Magnetic separation is based on the principle that different minerals have different magnetic properties. By subjecting a mixture of minerals to a magnetic field, it is possible to selectively attract and separate the desired minerals from the non-magnetic ones. This process is highly efficient and enables the extraction of valuable minerals that would otherwise be challenging to recover.

The magnetic separation process involves the use of a magnetic separator, a device that generates a magnetic field to attract and separate the minerals. The separator consists of a magnet, usually an electromagnet, and a rotating drum or conveyor belt. As the mixture of minerals pass through the magnetic field, the magnetic particles are attracted to the magnet or drum, while the non-magnetic particles continue their path. The separated minerals can then be collected and processed further.

The applications of magnetic separation in geological engineering are vast. It is extensively used in the concentration of minerals such as iron ore, chromite, and magnetite. Magnetic separation is also employed in the beneficiation of rare earth elements and the recovery of valuable metals from electronic waste. Furthermore, it plays a crucial role in coal cleaning processes, where unwanted minerals and impurities are removed to improve the quality of the final product.

The advantages of magnetic separation are numerous. It is a non-destructive process that does not require the use of chemicals, making it environmentally friendly. It is also highly efficient and can achieve high levels of mineral recovery. Additionally, magnetic separation is a cost-effective method that can be easily incorporated into existing mining operations.

In conclusion, magnetic separation is a vital technique in the field of resource extraction. Its ability to selectively separate minerals based on their magnetic properties makes it an invaluable tool for geological engineers. By harnessing the power of magnetic separation, mining operations can optimize mineral recovery, reduce waste, and contribute to sustainable resource extraction practices.

Flotation

Flotation is a crucial process in the field of geological engineering, serving as a key technique for the extraction of valuable resources from ore deposits. This subchapter will delve into the principles, methods, and innovations behind flotation, providing a comprehensive understanding of its importance and applications.

Flotation is a separation process that utilizes the differences in the surface properties of minerals to separate them from gangue materials. It is widely used in the mining industry to extract minerals such as copper, gold, and silver from their respective ores. The process involves the addition of specific chemicals, known as flotation reagents, to the ore pulp, which creates a froth that carries the desired minerals to the surface.

This subchapter will explore the fundamental principles that underpin flotation, including the theory of attachment and detachment of particles to and from air bubbles. It will also highlight the importance of understanding the mineralogical composition of ore deposits, as this knowledge is crucial for designing an effective flotation process.

Technological advancements have revolutionized flotation over the years, enabling more efficient and effective mineral recovery. This subchapter will discuss some of these innovations, such as the development of advanced flotation machines and the use of novel reagents. It will also touch upon the concept of selective flotation, which allows for the recovery of specific minerals while minimizing the loss of valuable resources.

Furthermore, this subchapter will address the environmental considerations associated with flotation, as the mining industry strives to minimize its ecological footprint. The utilization of sustainable reagents and the adoption of water recycling systems are some of the innovative approaches that will be explored.

Overall, this subchapter on flotation aims to provide a comprehensive understanding of this vital technique in geological engineering. It is essential reading for professionals in the mining industry, researchers, and students in the field of geological engineering. By mastering the principles and innovations of flotation, individuals will be better equipped to optimize mineral recovery, enhance resource extraction processes, and contribute to the sustainable development of the mining industry.

Chemical Extraction Techniques

Chemical extraction techniques play a crucial role in the field of geological engineering, allowing for the efficient and sustainable extraction of valuable resources from the Earth. In this subchapter, we will explore the various methods and innovations used in chemical extraction, providing an overview of the techniques employed by geological engineers to unlock the hidden potential of mineral deposits.

One of the most widely used chemical extraction techniques is leaching, which involves the use of solvents to dissolve valuable minerals from ores. This process is particularly effective in extracting metals such as gold, silver, and copper. Geological engineers carefully design leaching systems, considering factors such as the composition of the ore, the choice of leaching agent, and the optimal conditions for maximum efficiency. This subchapter will delve into the different types of leaching techniques, including heap leaching, in-situ leaching, and agitation leaching, discussing their advantages, limitations, and environmental considerations.

Another important chemical extraction technique discussed in this subchapter is solvent extraction. This method is commonly used to separate and recover specific metals from solution, offering a highly efficient and selective approach. Solvent extraction is employed in various stages of the mining process, including the recovery of valuable metals from pregnant leach solutions and the purification of metal concentrates. We will explore the principles and applications of solvent extraction, highlighting its significance in maximizing resource recovery.

Furthermore, this subchapter will touch upon the advancements in chemical extraction techniques, such as bioleaching and phytomining. Bioleaching involves the use of microorganisms to extract metals from low-grade ores, offering a more environmentally friendly and cost-effective alternative to traditional extraction methods. Phytomining, on the other hand, utilizes plants to absorb metals from the soil, providing a sustainable approach to resource extraction while simultaneously rehabilitating degraded land.

Throughout this subchapter, we will emphasize the importance of sustainable practices in chemical extraction. Geological engineers are increasingly focused on minimizing the environmental impact of mining operations, and the adoption of cleaner and more efficient extraction techniques is a crucial step in achieving this goal. By optimizing the use of chemicals, reducing waste generation, and implementing effective environmental management strategies, geological engineers can ensure the responsible extraction of resources while preserving the integrity of the surrounding ecosystems.

Whether you are a student, a professional in the field of geological engineering, or simply curious about the innovations in resource extraction, this subchapter on chemical extraction techniques will provide valuable insights into the advancements and challenges faced by the industry.

Hydrometallurgical Processes

Hydrometallurgical processes play a key role in the field of mining engineering, particularly in the extraction of valuable metals from ore. This subchapter explores the various techniques and innovations associated with these processes, shedding light on the advancements that have revolutionized the industry.

Hydrometallurgy involves the use of aqueous solutions to extract metals from their ores. It is a versatile and efficient method that can be applied to a wide range of minerals, including copper, gold, silver, and uranium, among others. The main advantage of hydrometallurgy lies in its ability to extract metals from low-grade ores that are otherwise uneconomical to process using traditional methods.

One of the most commonly used hydrometallurgical processes is leaching. This process involves the dissolution of metals from ore by using a leaching agent, such as sulfuric acid or cyanide, in a controlled environment. Leaching can be performed through various methods, including heap leaching, tank leaching, and in-situ leaching, each tailored to the specific characteristics of the ore deposit.

Another important aspect of hydrometallurgical processes is the extraction of metals from leach solutions. This is typically achieved through techniques like solvent extraction, precipitation, and electrowinning. Solvent extraction involves the use of organic compounds to selectively extract the desired metal ions from the leach solution, while precipitation involves the addition of reagents to induce the formation of solid metal compounds. Electro-winning, on the other hand, involves the use of an electrical current to deposit the metal ions onto a cathode.

Over the years, significant innovations have been made in the field of hydrometallurgy. Advances in process chemistry, metallurgical understanding, and equipment design have led to improved efficiency and reduced environmental impact. For instance, the use of biotechnology, such as bioleaching, has emerged as a sustainable alternative to traditional hydrometallurgical processes. Bioleaching utilizes microorganisms to extract metals from ores, offering a greener and more cost-effective solution.

In conclusion, hydrometallurgical processes have revolutionized the field of mining engineering. These techniques have enabled the extraction of valuable metals from low-grade ores, opening up new possibilities for resource extraction. With ongoing innovations and advancements, the future of hydrometallurgy looks promising, promising a more sustainable and efficient approach to metal extraction.

Pyrometallurgical Processes

Pyrometallurgical processes are a fundamental aspect of mining engineering that involve the use of high temperatures to extract valuable metals and minerals from their ores. This subchapter explores the techniques and innovations in pyrometallurgy, highlighting its significance in the field of geological engineering.

Introduction to Pyrometallurgical Processes

Pyrometallurgical processes have been employed for centuries to extract metals from ores. These processes utilize the heat generated by combustion reactions to separate valuable metals from their ores and produce concentrates that can be further processed. The subchapter dives into the various pyrometallurgical techniques and their applications in modern mining operations.

Smelting

Smelting is one of the most common pyrometallurgical processes used in mining engineering. It involves heating the ore to high temperatures in a furnace, which causes the metal to melt and separate from the impurities. The molten metal is then collected and further processed into a refined form. The subchapter provides an in-depth exploration of different smelting techniques, such as flash smelting and electric arc furnaces, and discusses their advantages and limitations.

Roasting

Roasting is another pyrometallurgical process that involves heating the ore in the presence of oxygen to remove volatile impurities and convert the metal sulfides into metal oxides. This process is particularly useful for the extraction of non-ferrous metals like copper, lead, and zinc. The subchapter explains the principles behind roasting and discusses the different types of roasting furnaces and their applications.

Reverberatory Furnaces

Reverberatory furnaces are commonly used in pyrometallurgical processes due to their versatility and efficiency. These furnaces utilize a combination of direct and indirect heat transfer to achieve the desired reactions. The subchapter explores the working principles of reverberatory furnaces and their applications in various pyrometallurgical processes.

Innovation in Pyrometallurgy

The subchapter also emphasizes the importance of innovation in pyrometallurgical processes. It highlights recent advancements and emerging technologies in the field, such as the use of plasma technology, microwave heating, and fluidized bed reactors. These innovations aim to improve energy efficiency, reduce environmental impact, and enhance the overall performance of pyrometallurgical processes.

Conclusion

Pyrometallurgical processes play a vital role in the extraction of valuable metals and minerals from ores. This subchapter provides a comprehensive overview of the techniques and innovations in pyrometallurgy, catering to the audience of geological engineering. Understanding these processes is essential for mining engineers to optimize resource extraction and meet the growing demands of the mining industry in a sustainable and efficient manner.

Chapter 6: Innovations in Mining Engineering

Automation and Robotics in Mining Operations

In recent years, advancements in technology have revolutionized the mining industry, particularly with the integration of automation and robotics into mining operations. These cutting-edge technologies have not only enhanced productivity but also improved safety and efficiency in resource extraction. This subchapter explores the various applications of automation and robotics in mining operations, providing valuable insights for geological engineering professionals and enthusiasts alike.

Automation has significantly transformed the mining landscape, enabling companies to streamline their processes and operate more efficiently. With automated systems, mining operations can be remotely controlled, reducing the need for human presence in hazardous environments. This not only enhances the safety of workers but also allows for continuous operations without interruptions. Furthermore, automation allows for real-time monitoring of equipment, ensuring timely maintenance and repairs, thus minimizing downtime and maximizing productivity.

Robotics, on the other hand, has opened up new avenues for mining operations by introducing the use of intelligent machines. These robots are designed to perform complex tasks that are dangerous or challenging for humans. For example, autonomous drilling rigs can precisely and efficiently extract resources from deep within the Earth, reducing human error and increasing accuracy. Additionally, robots equipped with high-resolution cameras and sensors can explore inaccessible areas, providing valuable geological data for accurate resource estimation.

The integration of automation and robotics in mining operations has not only improved efficiency but also reduced the environmental impact of resource extraction. Automated systems can optimize the use of resources, such as fuel and energy, resulting in lower emissions and decreased waste. Furthermore, robotics can be utilized in rehabilitation efforts to restore mining sites to their natural state, mitigating the environmental footprint of mining operations.

For geological engineering professionals, understanding and embracing automation and robotics is crucial to stay ahead in the rapidly evolving mining industry. By harnessing the power of these technologies, engineers can optimize mining processes, minimize risks, and maximize resource extraction. As the industry continues to advance, it is essential to keep abreast of the latest developments in automation and robotics, as they hold the key to a sustainable and efficient future in mining.

In conclusion, automation and robotics have revolutionized mining operations, offering enhanced safety, efficiency, and sustainability. This subchapter has provided a comprehensive overview of the applications of automation and robotics in the mining industry, catering to the interests of geological engineering professionals and enthusiasts. By embracing these technologies, the mining industry can unlock new possibilities and shape a more sustainable future for resource extraction.

Advanced Geological and Geotechnical Monitoring

In the field of geological engineering, advanced monitoring techniques play a crucial role in ensuring the safety and efficiency of resource extraction activities. With the constant evolution of technology, monitoring tools and methodologies have advanced significantly, enabling engineers to gain a deeper understanding of the earth's subsurface and mitigate potential risks associated with mining operations. This subchapter explores the various advanced geological and geotechnical monitoring techniques that are revolutionizing the field of geological engineering.

One of the key advancements in monitoring technology is the use of remote sensing techniques. Drones equipped with high-resolution cameras and LiDAR sensors can capture detailed images and data of mining sites, providing engineers with valuable information about the topography, geological structures, and potential hazards. This helps in creating accurate 3D models of the mining area, aiding in efficient design and planning of excavation activities.

Furthermore, the use of satellite imagery and ground-based radar systems allows continuous monitoring of ground movements and subsurface deformations. These techniques help detect any potential instability, such as landslides or ground subsidence, providing early warnings and allowing engineers to take preventive measures. Real-time monitoring systems, integrated with sensors and geotechnical instruments, provide constant data on variables such as ground vibrations, water pressure, and rock stress, enabling engineers to assess the stability of mining structures and ensure worker safety.

Advancements in geophysical imaging techniques have also greatly improved the understanding of subsurface geology. Seismic surveys, electrical resistivity tomography, and ground-penetrating radar are some of the techniques used to map the geological formations and identify potential mineral deposits. This knowledge helps in efficient resource extraction and reduces the chances of encountering unexpected geological complexities.

Moreover, the integration of data from various monitoring techniques into a centralized database allows for real-time analysis and decision-making. Advanced data analytics and machine learning algorithms help in identifying patterns and anomalies, enabling engineers to predict potential hazards and optimize mining operations accordingly.

In conclusion, advanced geological and geotechnical monitoring techniques are vital for ensuring the safety and efficiency of resource extraction activities. The constant evolution of technology has revolutionized the field of geological engineering, providing engineers with accurate and real-time information about mining sites. These advancements enable better planning, early detection of hazards, and improved resource extraction techniques. As the field continues to evolve, embracing and mastering these advanced monitoring techniques becomes essential for every geological engineer.

Sustainable Mining Practices

In recent years, there has been growing concern about the environmental impact of mining activities. As the global demand for minerals and resources continues to rise, it is imperative that the mining industry adopts sustainable practices to minimize its ecological footprint. This subchapter explores the concept of sustainable mining and highlights various techniques and innovations that can be employed by geological engineers to ensure responsible resource extraction.

Sustainable mining refers to the process of extracting minerals and resources in a way that minimizes environmental damage and maximizes social and economic benefits. It involves integrating environmental, social, and economic considerations into every stage of the mining lifecycle, from exploration to closure.

One key aspect of sustainable mining is reducing the use of harmful chemicals and minimizing water and energy consumption. For example, advancements in extraction techniques, such as bioleaching and phytomining, can help reduce the need for toxic chemicals and protect surrounding ecosystems. Geological engineers play a crucial role in implementing these innovative techniques and ensuring their efficiency.

Another important aspect is land reclamation and rehabilitation. Once mining activities are complete, it is essential to restore the land to its original state or convert it into a beneficial post-mining land use. This can include reforestation, creating wildlife habitats, or even repurposing the land for renewable energy projects. Geological engineers can contribute by designing effective reclamation plans and monitoring the progress to ensure successful restoration.

Furthermore, the concept of sustainable mining extends beyond environmental considerations. It also encompasses social responsibility, including the health and safety of workers, community engagement, and the promotion of local economic development. Geological engineers can facilitate community partnerships, implement safety measures, and ensure fair labor practices to fulfill these social obligations.

Technological advancements have played a significant role in promoting sustainable mining practices. Automation and robotics have improved operational efficiency while reducing the risk of accidents. Digitalization and data analytics enable real-time monitoring of environmental impacts, allowing for timely interventions to prevent any adverse effects.

Ultimately, sustainable mining practices are essential for the long-term viability of the mining industry. By integrating environmental, social, and economic considerations, geological engineers can ensure responsible resource extraction that benefits both present and future generations. Whether it is through adopting innovative technologies, implementing efficient extraction techniques, or prioritizing community engagement, the geological engineering community plays a pivotal role in shaping a sustainable future for the mining industry.

Digitalization and Data Analytics in Mining

In recent years, the mining industry has undergone a significant transformation due to the advent of digitalization and data analytics. These technological advancements have revolutionized the way geological engineering is approached, allowing for more efficient and sustainable resource extraction. This subchapter explores the role of digitalization and data analytics in mining, providing invaluable insights for both professionals and enthusiasts in the field of geological engineering.

Digitalization has paved the way for a more streamlined and interconnected mining industry. With the integration of sensors, Internet of Things (IoT) devices, and advanced data collection systems, mining operations now have access to real-time data from various stages of the extraction process. This data encompasses parameters such as geological formations, equipment performance, environmental conditions, and safety indicators. By leveraging this wealth of information, geological engineers can make informed decisions, optimize processes, and enhance productivity.

Data analytics plays a crucial role in transforming raw data into actionable insights. Through advanced algorithms and machine learning techniques, massive amounts of data can be analyzed to identify patterns, trends, and anomalies. This enables geological engineers to gain a deeper understanding of the geological structure, mineral deposits, and potential risks associated with mining operations. Moreover, data analytics can aid in predicting equipment failures, optimizing energy consumption, and minimizing environmental impact.

The benefits of digitalization and data analytics in mining are multi-fold. Firstly, these technologies improve safety by providing real-time monitoring of hazardous conditions. Early detection of potential dangers allows for timely interventions and reduces the risk of accidents. Secondly, digitalization and data analytics enable better resource management. By accurately assessing the quality and quantity of deposits, geological engineers can optimize extraction techniques and minimize waste.

Furthermore, these advancements contribute to sustainability in the mining industry. By analyzing environmental data, such as air quality and water usage, mining operations can implement measures to reduce their ecological footprint. Additionally, digitalization facilitates remote monitoring and control of equipment, reducing the need for on-site personnel and minimizing the industry's carbon footprint.

In conclusion, digitalization and data analytics have revolutionized the field of geological engineering in the mining industry. These technologies have enhanced productivity, safety, and sustainability, making resource extraction more efficient and responsible. Aspiring geological engineers and professionals in the field can benefit greatly from understanding and harnessing the power of digitalization and data analytics in their pursuit of mastering mining engineering.

Chapter 7: Environmental and Social Considerations in Mining

Environmental Impact Assessment

Environmental Impact Assessment (EIA) is a crucial component of the mining industry that aims to evaluate the potential environmental consequences of resource extraction projects. This subchapter explores the significance of EIA in geological engineering, shedding light on its purpose, methods, and benefits.

In the field of geological engineering, EIA plays a pivotal role in identifying and assessing the environmental impacts of mining activities. The primary objective of an EIA is to ensure sustainable development by promoting responsible mining practices that minimize adverse effects on the environment. By conducting a comprehensive EIA, geological engineers can understand the potential environmental risks associated with mining operations and develop strategies to mitigate them effectively.

The process of conducting an EIA involves various steps, starting with scoping and baseline studies, followed by impact prediction, evaluation, and finally, the formulation of mitigation measures. Geological engineers work closely with environmental scientists and other stakeholders to gather data on key environmental parameters such as air and water quality, biodiversity, and soil stability. This data is then analyzed to predict potential impacts, such as habitat destruction, water contamination, and air pollution.

The evaluation of impacts is a critical step in the EIA process. It enables geological engineers to determine the significance of potential environmental effects, considering factors such as the scale, duration, and reversibility of impacts. This evaluation provides valuable insights for decision-making, aiding in the optimization of mining operations and the reduction of harmful consequences.

One of the key benefits of EIA in geological engineering is its ability to facilitate sustainable mining practices. By identifying potential environmental risks, geological engineers can implement measures to minimize impacts and ensure the preservation of ecosystems. These measures may include the implementation of advanced waste management systems, the use of eco-friendly mining technologies, and the reclamation of disturbed land.

Furthermore, EIA also promotes transparency and public participation in the decision-making process. It allows for the involvement of local communities, indigenous peoples, and other stakeholders, providing them with a platform to voice their concerns and opinions. This inclusive approach not only enhances the credibility of mining projects but also fosters a sense of responsibility and accountability among mining companies.

In conclusion, Environmental Impact Assessment is an essential tool in the field of geological engineering. By evaluating potential environmental impacts, geological engineers can adopt sustainable mining practices that prioritize environmental preservation. The implementation of EIA ensures the responsible extraction of resources while considering the needs of local communities and safeguarding ecosystems for future generations.

Rehabilitation and Mine Closure Planning

In the realm of geological engineering, the process of resource extraction is a critical component of the industry. However, it is equally vital to focus on the rehabilitation and mine closure planning to ensure the sustainable and responsible development of mining projects. This subchapter delves into the importance of rehabilitation and mine closure planning, exploring the techniques and innovations that can be employed in this process.

The primary objective of mine closure planning is to restore the land and surrounding environment to a state that is safe, stable, and environmentally sustainable. This involves minimizing the potential negative impacts on the ecosystem, water resources, and local communities. By implementing effective rehabilitation strategies, mining sites can be transformed into productive and beneficial landscapes.

Rehabilitation practices can vary depending on the specific characteristics of the mine, including its location, geology, and the type of mineral being extracted. Techniques such as re-vegetation, soil stabilization, and water management are commonly employed to restore the ecosystem and enhance biodiversity. Advances in technology have also allowed for the use of innovative methods, including the integration of remote sensing and GIS mapping, to monitor and assess the effectiveness of rehabilitation efforts.

Furthermore, mine closure planning extends beyond the physical aspects of the site. Social and economic considerations are also essential components of this process. Engaging with local communities and stakeholders from the early stages of a mining project is crucial to understanding their needs and concerns.

Establishing partnerships and implementing sustainable development initiatives can help create a positive legacy for the community.

This subchapter will explore case studies and best practices from around the world, highlighting successful examples of mine closure planning in action. It will showcase the innovative techniques and technologies that have been utilized to ensure effective rehabilitation, environmental protection, and community engagement. Through this comprehensive exploration, readers will gain a deeper understanding of the importance of rehabilitation and mine closure planning in the field of geological engineering.

For geological engineers, this subchapter will serve as a valuable resource, providing insights into the latest techniques and innovations in mine closure planning. It will equip professionals with the knowledge and skills needed to contribute to sustainable mining practices and responsible resource extraction. Additionally, it will offer guidance for mining companies, regulators, and policymakers to ensure that mining operations are conducted in a manner that prioritizes environmental stewardship and community well-being.

Community Engagement and Social Responsibility

In the field of geological engineering, community engagement and social responsibility play a vital role in ensuring sustainable and responsible resource extraction practices. As mining projects have the potential to significantly impact local communities and the environment, it is crucial for mining engineers to actively engage with the community and uphold their social responsibilities.

Community engagement involves establishing meaningful relationships with the local community and stakeholders, including indigenous groups, landowners, and residents. By understanding and addressing their concerns, mining engineers can build trust and foster cooperation, leading to mutually beneficial outcomes for all parties involved. This engagement should begin at the early stages of a mining project and continue throughout its lifecycle.

One of the key aspects of community engagement is effective communication. Mining engineers should communicate transparently and openly with the community, providing them with accurate information about the project's objectives, potential impacts, and mitigation measures. Regular community meetings, public consultations, and information sessions can facilitate this communication process, allowing residents to voice their concerns and be actively involved in decision-making processes.

Furthermore, social responsibility goes beyond merely engaging with the community and extends to implementing initiatives that contribute to the social and economic development of the local area. Mining engineers should strive to create employment opportunities for local residents, promote education and training programs, and support local businesses and infrastructure development. By investing in the well-being of the community, mining projects can leave a positive legacy and contribute to sustainable development.

Additionally, environmental responsibility is a crucial aspect of social responsibility in mining engineering. Engineers should implement environmentally conscious practices, such as minimizing the use of natural resources, reducing waste generation, and implementing effective rehabilitation and reclamation measures. By mitigating and minimizing the environmental impacts of mining operations, engineers can ensure the long-term sustainability of the ecosystem and protect biodiversity.

In conclusion, community engagement and social responsibility are essential components of responsible resource extraction in geological engineering. By actively engaging with the local community, addressing their concerns, and upholding social responsibilities, mining engineers can ensure sustainable development and leave a positive impact on the communities and environments in which they operate.

Chapter 8: Case Studies in Mining Engineering

Case Study 1: Successful Implementation of Advanced Mining Techniques

Mining plays a crucial role in the extraction of valuable resources from the Earth, and it is essential to constantly explore and adopt innovative techniques to optimize efficiency and sustainability. In this case study, we delve into the successful implementation of advanced mining techniques that have revolutionized the field of geological engineering.

Introduction:
With the rising demand for minerals and metals, traditional mining methods have become inadequate to meet the industry's needs. However, advancements in technology and the application of cutting-edge techniques have paved the way for a new era in mining engineering. This case study highlights a real-life success story where advanced mining techniques have been employed to achieve outstanding results.

Challenges   Faced:
The mining project under consideration was located in a complex geological region, posing significant challenges to extraction operations. The presence of intricate mineral deposits, varying geological structures, and environmental concerns demanded an innovative approach to overcome these obstacles.

Innovative   Solutions:
To tackle the geological complexities, the project team utilized state-of-the-art geological mapping techniques, including 3D modeling and remote sensing. These technologies allowed for a detailed

understanding of the deposit's characteristics, enabling precise resource estimation and optimal mine planning.

Furthermore, advanced drilling and blasting techniques were adopted to enhance productivity while minimizing environmental impact. The implementation of precision drilling, utilizing automated drilling rigs and advanced drilling patterns, ensured accurate fragmentation and reduced energy consumption.

Automation and Robotics: Another critical aspect of the successful implementation was the integration of automation and robotics in the mining process. Robotic systems were employed for tasks such as ore transportation, underground mapping, and safety inspections. These technologies not only improved efficiency but also enhanced worker safety by minimizing exposure to hazardous conditions.

Environmental Sustainability: The project also prioritized environmental sustainability by implementing advanced waste management techniques. By utilizing environmentally friendly tailings disposal methods and adopting stringent monitoring practices, the project significantly reduced its ecological footprint.

Results and Impact: The implementation of advanced mining techniques resulted in substantial benefits for both the project and the industry as a whole. The project achieved increased productivity, reduced operational costs, and enhanced safety standards. Moreover, the successful implementation served as a benchmark for other mining operations, inspiring the industry to embrace technological advancements and adopt sustainable practices.

Conclusion:

This case study exemplifies the power of advanced mining techniques in overcoming geological complexities and achieving remarkable success. By leveraging innovative technologies, mining engineers can optimize resource extraction while minimizing environmental impact. The lessons learned from this case study can be applied across the field of geological engineering, paving the way for a more sustainable and efficient mining industry.

Case Study 2: Overcoming Environmental Challenges in Mining Projects

Introduction:
Mining projects play a crucial role in meeting the growing demand for natural resources worldwide. However, these projects often face significant environmental challenges that must be addressed to ensure sustainable resource extraction. In this case study, we will explore the innovative techniques and strategies employed by geological engineers to overcome environmental challenges in mining projects.

The Importance of Environmental Considerations in Mining: Mining projects have historically been associated with environmental degradation, including deforestation, soil erosion, water pollution, and habitat destruction. Recognizing the importance of environmental sustainability, modern mining practices now prioritize minimizing these negative impacts. Geological engineers play a vital role in developing innovative solutions that ensure responsible mining operations.

Case   Study   Overview:
In this case study, we will examine a specific mining project that faced significant environmental challenges and the techniques used to overcome them. By understanding the strategies employed by geological engineers, we can gain insights into sustainable mining practices that benefit both the environment and local communities.

Addressing   Water   Pollution:
One of the major environmental challenges in mining projects is water pollution caused by the release of harmful chemicals and heavy metals. Geological engineers have developed innovative methods to mitigate this issue, such as the implementation of advanced water treatment

technologies and the use of chemical additives to neutralize pollutants. These techniques effectively reduce the impact on water bodies and surrounding ecosystems.

Restoring   Biodiversity:
Mining projects often result in habitat destruction and loss of biodiversity. Geological engineers work closely with environmental scientists to develop effective reclamation and restoration plans. By replanting native vegetation, creating artificial habitats, and employing ecological engineering techniques, mining sites can be transformed into thriving ecosystems once mining operations cease.

Mitigating   Air   Pollution:
Mining activities can lead to air pollution through dust emissions and the release of harmful gases. Geological engineers employ various strategies to address this challenge, including the use of dust suppression systems, proper ventilation, and the implementation of emission control technologies. These measures ensure that air quality is maintained at acceptable levels, minimizing the impact on local communities and the environment.

Conclusion:
Effective management of environmental challenges is essential for the sustainable development of mining projects. Geological engineers play a crucial role in developing innovative techniques and strategies to overcome these challenges while ensuring responsible resource extraction. By implementing advanced technologies and adopting environmentally friendly practices, mining projects can coexist harmoniously with the environment and contribute to the welfare of local communities. This case study demonstrates the importance of

environmental considerations in mining and highlights the role of geological engineers in driving positive change in the industry.

Case Study 3: Sustainable Mining Practices and Benefits

Mining plays a vital role in our modern society, providing essential resources for various industries. However, the traditional mining methods have often come under scrutiny due to their negative environmental impacts. In recent years, the concept of sustainable mining has gained significant attention, focusing on minimizing these impacts and maximizing the benefits for both the industry and the environment. This case study explores sustainable mining practices and their many benefits.

Sustainable mining practices involve the integration of environmental, social, and economic considerations into every stage of the mining process. One key aspect is the adoption of advanced technologies and engineering techniques to reduce the environmental footprint. For example, the use of remote sensing and satellite imagery helps identify potential mining sites with minimal ecological disturbances. Additionally, implementing efficient waste management systems and reclamation plans ensures the restoration of the land post-mining activities.

Social responsibility is another critical component of sustainable mining. Engaging with local communities and indigenous populations ensures their participation and benefits from mining activities. This includes providing employment opportunities, promoting education and training, and supporting local businesses. By involving stakeholders in decision-making processes, sustainable mining fosters a sense of ownership and ensures the well-being of these communities.

Economically, sustainable mining practices have proven to be advantageous. By minimizing waste generation and optimizing resource utilization, mining operations become more cost-effective.

Moreover, sustainable practices enhance the industry's reputation, attracting responsible investors and improving access to capital. The implementation of innovative technologies also leads to increased productivity and efficiency, resulting in higher profitability in the long run.

In terms of environmental benefits, sustainable mining practices help mitigate the negative impacts on air, water, and soil quality. Advanced techniques such as water recycling, dust suppression, and tailings management minimize pollution and prevent the release of harmful substances. Additionally, sustainable mining promotes the use of renewable energy sources, reducing greenhouse gas emissions and combating climate change.

For geological engineers, understanding and implementing sustainable mining practices is crucial. It not only ensures the longevity and profitability of mining projects but also contributes to the preservation of Earth's resources. By incorporating environmental considerations into their designs, geological engineers can minimize the ecological footprint of mining operations and optimize resource extraction.

In conclusion, sustainable mining practices offer a holistic approach to resource extraction, addressing environmental, social, and economic concerns. By adopting advanced technologies, engaging with local communities, and minimizing environmental impacts, the mining industry can reap long-term benefits. Geological engineers play a vital role in implementing these practices, ensuring responsible and sustainable mining operations for future generations.

Chapter 9: Future Trends and Emerging Technologies in Mining Engineering

Exploration Technologies for Deeper and Remote Deposits

In the field of geological engineering, the search for valuable mineral resources often takes us to deeper and more remote deposits. These untapped resources hold immense potential, but their extraction poses unique challenges that require innovative exploration technologies. This subchapter delves into the techniques and innovations that are revolutionizing resource extraction in these deeper and more remote deposits.

One of the key technologies driving exploration in deeper deposits is remote sensing. Remote sensing techniques, involving the use of satellites and airborne sensors, allow geological engineers to gather valuable data without physically being present at the site. Through the analysis of electromagnetic waves and other signals, these technologies can provide detailed information about the geological composition, mineral content, and structural characteristics of the deposits. This helps in determining the feasibility of mining operations and reduces the need for costly and time-consuming physical exploration.

Furthermore, advancements in seismic imaging have greatly enhanced our ability to explore deeper deposits. By utilizing sophisticated seismic techniques, geological engineers can create detailed 3D models of the subsurface, enabling them to accurately locate and delineate mineral deposits. These models provide valuable insights into the geological structures and help optimize the extraction process, reducing the risk of costly errors.

Another important area of exploration technology for deeper and remote deposits is the use of autonomous drones and robots. These unmanned systems can access areas that are difficult or dangerous for humans to reach, providing valuable geological data and imagery. Equipped with advanced sensors and cameras, these autonomous devices can gather high-resolution data and create detailed maps of the deposits, aiding in the planning and execution of mining operations.

Furthermore, the application of artificial intelligence (AI) and machine learning algorithms has revolutionized the processing and interpretation of exploration data. These technologies can quickly analyze vast amounts of geological and geophysical data, identifying patterns and anomalies that would be difficult for human experts to detect. By leveraging AI, geological engineers can make more informed decisions about the locations and methods of resource extraction, optimizing productivity and minimizing environmental impact.

In conclusion, the exploration technologies for deeper and remote deposits have transformed the field of geological engineering. Remote sensing, seismic imaging, autonomous drones, and AI-driven data analysis have revolutionized the way we explore and extract resources. These innovations not only enhance our ability to discover new deposits but also improve the efficiency and sustainability of mining operations. As the demand for mineral resources continues to grow, mastering these exploration technologies is essential for geological engineers to unlock the full potential of deeper and remote deposits.

Alternative Energy Sources for Mining Operations

Mining operations have long been associated with high energy consumption and environmental impact. However, in recent years, the industry has been actively exploring alternative energy sources to reduce its carbon footprint and improve sustainability. This subchapter aims to introduce various alternative energy sources that can be utilized in mining operations, catering to the audience of geological engineering and anyone interested in sustainable resource extraction.

One of the most promising alternative energy sources for mining is solar power. With advancements in photovoltaic technology, solar panels are becoming more efficient and cost-effective. By harnessing the abundant sunlight available in many mining regions, solar power can be used to generate electricity for various mining processes, including lighting, ventilation, and even powering heavy machinery. Furthermore, solar energy can be stored in batteries to ensure a continuous power supply, even during cloudy days or at night.

Wind power is another viable option for mining operations, especially in areas with strong and consistent winds. Wind turbines can be installed to generate electricity, which can be integrated into the existing power grid or used on-site. Wind energy is renewable, emission-free, and has the potential to significantly reduce the reliance on fossil fuels in mining operations.

Hydropower, derived from flowing or falling water, is also a valuable alternative energy source for mining. Mining operations often require large quantities of water, and by utilizing hydropower, the energy generated from water can be harnessed to meet the energy demands of the operation. This can be achieved through the installation of hydroelectric turbines in waterways or by utilizing water pressure for mechanical power.

Geothermal energy, which harnesses heat from the Earth's core, holds great potential for mining operations situated in geologically active regions. By tapping into the Earth's natural heat, geothermal power can provide a constant and reliable source of energy for mining processes. Additionally, geothermal energy can be used for heating or cooling purposes, further reducing the environmental impact of mining operations.

In conclusion, alternative energy sources offer a promising solution to reduce the environmental impact of mining operations while ensuring a sustainable future for resource extraction. Solar power, wind power, hydropower, and geothermal energy are all viable options that can be integrated into mining operations, benefiting both the industry and the environment. By embracing these alternative energy sources, the mining industry can make significant strides towards a greener and more sustainable future.

Integration of Artificial Intelligence in Mining Processes

Artificial Intelligence (AI) has revolutionized various industries, and mining is no exception. In recent years, the integration of AI in mining processes has emerged as a game-changer, improving efficiency, safety, and sustainability in the field of geological engineering. This subchapter explores the incredible potential of AI in transforming mining operations and the benefits it offers to the industry as a whole.

AI-powered technologies have the ability to process and analyze massive amounts of data with unparalleled speed and accuracy. In the context of geological engineering, this means that AI can assist in identifying and characterizing mineral deposits, optimizing drilling and blasting techniques, and enhancing the overall decision-making process. By leveraging machine learning algorithms, AI systems can learn from historical data and make predictions, allowing mining companies to make informed decisions based on real-time information.

One of the key areas where AI is making a significant impact is in exploration and prospecting. Traditional exploration methods are time-consuming and costly, often involving extensive drilling and sampling. AI algorithms can analyze geological and geophysical data to identify patterns and anomalies that indicate the presence of valuable minerals. This enables companies to prioritize exploration efforts and target areas with the highest potential for resource extraction, saving both time and resources.

Moreover, AI can greatly improve safety in mining operations by automating dangerous tasks and minimizing human exposure to hazardous environments. Robotics and autonomous vehicles equipped with AI algorithms can be deployed in underground mines, reducing the risk of accidents and ensuring the safety of workers. Additionally, AI can monitor and analyze real-time data from various sensors, detecting potential dangers such as gas leaks or ground instability, and alerting the mining personnel for timely action.

Furthermore, the integration of AI in mining processes has significant environmental benefits. By optimizing resource extraction techniques, AI can reduce the environmental footprint of mining operations. For instance, AI algorithms can optimize the use of energy and water resources, minimize waste generation, and decrease the carbon emissions associated with mining activities.

In conclusion, the integration of AI in mining processes is transforming the field of geological engineering. By leveraging AI-powered technologies, mining companies can enhance exploration efforts, improve safety, and reduce the environmental impact of their operations. As the industry embraces AI, it is crucial for geological engineers to understand and master these innovative techniques to stay at the forefront of this technological revolution.

Chapter 10: Career Opportunities and Professional Development in Mining Engineering

Job Roles and Responsibilities in the Mining Industry

The mining industry plays a crucial role in the extraction and utilization of valuable resources from the Earth's crust. With the ever-increasing global demand for minerals and metals, it is essential to understand the various job roles and responsibilities within this industry. This subchapter aims to provide an overview of the diverse job opportunities available in the mining sector, specifically targeting individuals interested in geological engineering.

Geological engineering professionals are vital players in the mining industry, responsible for assessing the feasibility and economic viability of mineral deposits. They play a crucial role in locating and evaluating potential mining sites, conducting geological surveys, and determining the quality and quantity of mineral resources present. These experts utilize sophisticated technologies, such as remote sensing and geophysical surveys, to gather data and analyze geological formations accurately.

One of the key job roles in geological engineering is that of a geologist. Geologists specialize in studying the Earth's composition, structure, and processes to identify mineral deposits and assess their viability for commercial extraction. They conduct fieldwork, collect samples, and analyze geological data to provide valuable insights for mining companies. Geologists also play a vital role in ensuring environmental sustainability by minimizing the negative impacts of mining activities on the surrounding ecosystems.

Another crucial job role in the mining industry is that of a mining engineer. Mining engineers utilize their technical expertise to plan, design, and oversee mining operations. They are responsible for optimizing the extraction process, ensuring safety protocols, and maximizing resource recovery. Mining engineers collaborate with geologists and other professionals to develop efficient mining strategies that minimize costs and environmental impacts.

In addition to geologists and mining engineers, the mining industry offers a wide range of job roles, including environmental specialists, surveyors, safety officers, and equipment operators. Each of these roles contributes to the smooth functioning of mining operations and the overall sustainability of the industry.

It is important to note that the mining industry offers significant career opportunities for individuals with diverse skill sets, ranging from technical expertise to environmental stewardship. Aspiring geological engineering professionals can explore various educational and training programs to acquire the necessary knowledge and skills to excel in their chosen job roles within the mining industry.

In conclusion, the mining industry offers a multitude of job roles and responsibilities for individuals interested in geological engineering. From geologists to mining engineers and environmental specialists, each role contributes to the successful extraction of valuable resources while ensuring environmental sustainability. Aspiring professionals in this field can find rewarding careers in the mining industry by acquiring the necessary skills and knowledge through education and training programs.

Education and Training Pathways for Mining Engineers

Mining engineers play a crucial role in the field of resource extraction, as they are responsible for designing and implementing safe and efficient mining operations. If you have a passion for geological engineering and aspire to become a mining engineer, it is essential to understand the education and training pathways available to you.

To embark on a career in mining engineering, a bachelor's degree in mining engineering or a related field is typically required. This degree equips students with a solid foundation in core engineering disciplines such as mathematics, physics, geology, and computer science. It also provides specialized knowledge in areas specific to mining engineering, such as mine planning, mineral processing, and mine safety.

Many universities offer bachelor's programs in mining engineering, allowing students to gain hands-on experience through fieldwork and internships. These practical experiences are invaluable, as they provide students with real-world exposure to the challenges and complexities of the mining industry.

Upon completing a bachelor's degree, aspiring mining engineers can choose to further enhance their knowledge and skills by pursuing a master's degree in mining engineering. A master's degree offers a deeper understanding of advanced mining technologies, mine management, and sustainable resource development. It also provides opportunities for research and specialization in specific areas of interest, such as mine ventilation, mine automation, or environmental impact assessment.

Beyond formal education, mining engineers benefit from continuous professional development and on-the-job training. The mining industry is constantly evolving, with new technologies and practices emerging regularly. Therefore, staying up-to-date with the latest developments is crucial for mining engineers to excel in their careers.

Professional organizations, such as the Society for Mining, Metallurgy, and Exploration (SME), offer various resources, conferences, and workshops to support ongoing education and training for mining engineers. These opportunities allow engineers to network with industry professionals, exchange knowledge, and stay informed about the latest advancements and best practices in the field.

In conclusion, a career in mining engineering requires a solid educational foundation and continuous professional development. By pursuing a bachelor's degree in mining engineering, followed by a master's degree and engaging in ongoing training and education, aspiring mining engineers can acquire the necessary skills and knowledge to thrive in this dynamic industry. Whether your interest lies in geological engineering or any related field, embarking on this educational and training pathway can open doors to exciting opportunities and contribute to sustainable resource extraction practices.

Professional Organizations and Networking Opportunities

In any field, professional organizations play a vital role in fostering growth and development. This is especially true for the niche of Geological Engineering within the broader realm of mining engineering. These organizations provide an invaluable platform for individuals to connect, collaborate, and stay updated on the latest trends and innovations in their field.

One such professional organization that caters to the needs of Geological Engineers is the Society for Mining, Metallurgy, and Exploration (SME). SME brings together professionals from the mining industry, including geological engineers, to share knowledge, experiences, and best practices. Through their conferences, workshops, and technical sessions, SME offers numerous networking opportunities for geological engineers to connect with industry experts, researchers, and peers. Attending these events can prove to be a game-changer, as they provide a chance to learn from the best, gain insights into cutting-edge technologies, and establish valuable professional connections.

Another prominent professional organization that geological engineers can benefit from is the Geological Society of America (GSA). GSA aims to promote the understanding of Earth sciences and fosters professional growth among its members. The society organizes regular conferences, symposiums, and field trips that allow geological engineers to expand their knowledge base and interact with fellow professionals. Additionally, GSA offers various online platforms where members can participate in discussions, share research findings, and seek advice from seasoned experts in the field.

Networking opportunities provided by these professional organizations extend beyond conferences and events. They often have dedicated online forums, discussion boards, and social media groups where geological engineers can connect with like-minded individuals. These platforms serve as virtual meeting places for professionals to share ideas, ask questions, and seek guidance on challenges they face in their day-to-day work.

Joining professional organizations not only enhances one's technical expertise but also opens doors to career advancement. Many organizations offer mentorship programs that pair young professionals with experienced mentors who can guide them on their career path. Additionally, they often advertise job openings and provide access to exclusive job boards, increasing the chances of career growth and progression for geological engineers.

In conclusion, professional organizations and networking opportunities are crucial for geological engineers aiming to excel in their field. By becoming a member of such organizations, geological engineers can access a vast pool of knowledge, connect with industry experts, and enhance their professional network. The benefits of these associations extend not only to personal growth but also to career advancement, making them a must-have resource for every geological engineer striving for excellence.

Chapter 11: Conclusion and Final Remarks

Recap of Key Learnings

In the exciting and ever-evolving field of geological engineering, it is essential to continually learn and adapt to new techniques and innovations for resource extraction. Throughout this book, "Mastering Mining Engineering: Techniques and Innovations for Resource Extraction," we have explored a wide range of topics and shared valuable insights that are applicable to everyone interested in the field of geological engineering.

One of the key learnings we have emphasized is the importance of understanding the geological context. By comprehending the geological formations, structures, and properties, engineers can effectively plan and design resource extraction projects. From studying the characteristics of different rock types to analyzing fault lines and understanding the impact of tectonic movements, geological knowledge is the foundation of successful mining engineering.

Another crucial aspect we have covered is the significance of sustainable mining practices. As the world becomes more conscious of environmental impacts, it is essential for engineers to minimize the ecological footprint of mining operations. We have explored techniques such as reclamation, rehabilitation, and the use of advanced technologies to mitigate environmental damage and ensure the responsible extraction of resources.

Risk assessment and management have also been highlighted as vital components of mining engineering. From identifying potential hazards to implementing strategies to mitigate them, understanding and effectively managing risks is crucial for the safety of workers and the success of mining projects. We have discussed various methodologies, such as the use of advanced monitoring systems, geotechnical analysis, and safety protocols, to minimize risks in mining operations.

Additionally, our exploration of technological advancements has shown how innovations are transforming the field of geological engineering. From the use of drones for aerial surveys to the implementation of artificial intelligence and machine learning algorithms for data analysis, embracing and harnessing these technologies can greatly enhance the efficiency and accuracy of mining operations.

Lastly, we have emphasized the importance of collaboration and interdisciplinary approaches in geological engineering. By working together with geologists, environmental scientists, and other experts, engineers can benefit from diverse perspectives and integrate various disciplines to optimize resource extraction projects.

In conclusion, "Mastering Mining Engineering: Techniques and Innovations for Resource Extraction" has provided a comprehensive overview of key learnings in geological engineering. Whether you are a seasoned professional or just starting your journey in this field, understanding the geological context, practicing sustainable mining techniques, managing risks, embracing technological advancements, and fostering collaboration will be crucial for success in the fascinating world of geological engineering.

The Importance of Mining Engineering in Resource Extraction

Mining engineering plays a vital role in the extraction and utilization of Earth's valuable resources. This subchapter aims to shed light on the significance of mining engineering in resource extraction, specifically focusing on the niches of geological engineering. Whether you are an aspiring mining engineer, a geologist, or simply someone interested in learning about the intricacies of resource extraction, this subchapter will provide you with valuable insights.

Mining engineering encompasses a wide range of disciplines, including geology, geotechnical engineering, rock mechanics, and mineral processing. These fields work together to ensure the safe and efficient extraction of minerals, metals, and other valuable resources from the Earth's crust. Without mining engineering, the extraction and utilization of these resources would be challenging, if not impossible.

Geological engineering, in particular, plays a crucial role in mining operations. It involves the study of the Earth's structure, composition, and processes, focusing on identifying and assessing mineral deposits. Through geological engineering, mining engineers gain a deep understanding of the geological conditions and develop strategies to extract resources effectively.

One key aspect of mining engineering is the exploration and evaluation of mineral deposits. Geological engineers utilize various techniques, including remote sensing, geochemical analysis, and geological mapping, to identify potential resource-rich areas. This information helps mining companies make informed decisions about investing in exploration and development projects.

Moreover, mining engineering ensures the efficient extraction of resources while minimizing environmental impacts. Geological engineers assess the stability of mining sites, design safe mine layouts, and develop strategies for waste management and reclamation. By integrating sustainable practices, mining engineers strive to minimize the ecological footprint associated with resource extraction.

The importance of mining engineering extends beyond resource extraction. It contributes to economic growth and development by providing employment opportunities and generating revenue. Countries with abundant mineral resources often rely on mining industries as a significant source of income.

In conclusion, mining engineering, especially geological engineering, is of utmost importance in resource extraction. It combines various disciplines to ensure the safe and efficient extraction of minerals while minimizing environmental impacts. Understanding the significance of mining engineering is crucial for aspiring mining engineers, geologists, and anyone interested in the field of resource extraction.

Looking Forward: The Future of Mining Engineering

As we step into the future, the field of mining engineering is poised for significant advancements and innovations. The constant demand for natural resources, coupled with environmental concerns and technological advancements, will shape the future of this industry. In this subchapter, we will explore the key trends and developments that will define the future of mining engineering.

One of the most significant shifts in the field of mining engineering is the increased focus on sustainability and environmental responsibility. Mining companies are now recognizing the importance of minimizing their ecological footprint and are investing in advanced technologies to achieve this goal. From using renewable energy sources to implementing efficient waste management systems, the future of mining engineering will prioritize sustainable practices.

Technological advancements, such as automation and artificial intelligence (AI), are transforming the mining industry. Robotics and autonomous vehicles are being employed to increase operational efficiency and improve safety standards. AI-powered algorithms are aiding in data analysis, making mining operations more accurate and effective. The future of mining engineering will see the integration of these technologies to streamline processes and enhance productivity.

Moreover, the use of advanced geological modeling techniques will revolutionize resource extraction. Geological engineers will be able to accurately map and model underground deposits, allowing for more efficient extraction methods. Through the use of 3D modeling, real-time monitoring, and advanced data analytics, mining engineers will be able to optimize production and reduce costs.

The future of mining engineering will also witness a shift towards deep-sea mining. With the depletion of land-based resources, the exploration and extraction of minerals from the ocean floor will become crucial. Geological engineers specializing in underwater mining will play a pivotal role in developing technologies and techniques to extract valuable resources from the depths of the ocean.

Furthermore, the integration of virtual reality (VR) and augmented reality (AR) will transform the way mining engineers visualize and plan operations. VR and AR technologies will allow engineers to simulate mining scenarios, test different strategies, and identify potential risks before executing them in the real world. This will not only enhance safety but also optimize resource extraction.

In conclusion, the future of mining engineering holds tremendous potential for technological advancements and sustainable practices. As the demand for natural resources continues to grow, mining engineers will play a vital role in ensuring responsible extraction and minimizing environmental impact. By embracing advanced technologies, adopting sustainable practices, and exploring new frontiers such as deep-sea mining, mining engineering is set to thrive in the future.